Water Scientists

William B. Rice

Earth and Space Science Readers:
Water Scientists

Publishing Credits

Editorial Director
Dona Herweck Rice

Associate Editor
Joshua BishopRoby

Editor-in-Chief
Sharon Coan, M.S.Ed.

Creative Director
Lee Aucoin

Illustration Manager
Timothy J. Bradley

Publisher
Rachelle Cracchiolo, M.S.Ed.

Science Contributor
Sally Ride Science

Science Consultants
William B. Rice,
Engineering Geologist
Nancy McKeown,
Planetary Geologist

Teacher Created Materials
5301 Oceanus Drive
Huntington Beach, CA 92649-1030
http://www.tcmpub.com
ISBN 978-0-7439-0556-5

Reprinted 2012
BP 5028

Table of Contents

Wonderful Water

Drink it. Wash in it. Clean with it. Splash in it. Play in it. You can even work with it. What is it? Water, of course.

Water is the amazing substance that is so important we barely give it a thought. We use it all the time. We take it for granted. We don't really think about it until it isn't there.

Just stop for a minute and think about every time you have used water today. You have probably used water in some way more than you have done just about anything else. You clean with it and drink it just about every day. Water is also fun to play and swim in. It's beautiful to look at in oceans, lakes, and streams. It's also necessary for every living thing to keep on living.

Because of water's importance, people through time have made it their job to study water. They learn where water comes from, what it's made of, and what it does. They find ways to keep water clean and healthy. They figure out how to keep water flowing so everyone has what he or she needs. These people are water scientists. We couldn't get along without them.

A scientist who studies Earth is a geologist. A water scientist is a hydrologist.

What Is Water?

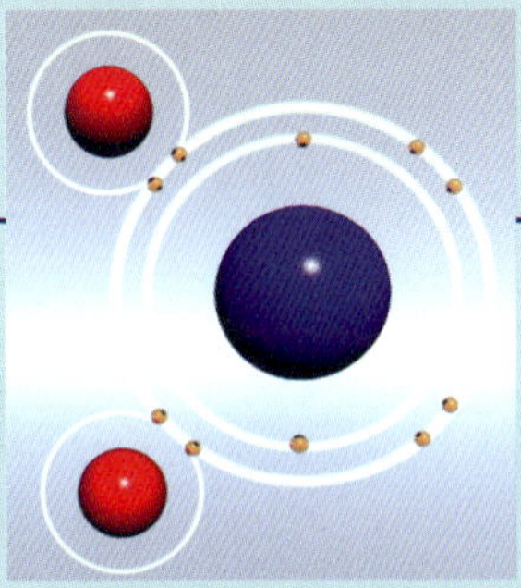

To make a water molecule, the hydrogen and oxygen atoms bond.

Water is made of two **elements**. An element is a basic substance that is naturally made. The elements that make water are hydrogen and oxygen. Two hydrogen **atoms** combine with one oxygen atom to make one **molecule** of water. It is written as H_2O.

Billions of water molecules combine to make the liquid called water. But, H_2O doesn't have to be just a liquid. It can also be a solid or a gas. Either way, it's still water. It changes form as it moves through the **water cycle**.

Mohammed Karaji

The eastern world has always been an important place of learning. This was especially true in the tenth century. It was a time of many new ideas and new achievements. In the Middle East, the studies of math and science were especially important. Scientists were honored. The people valued new knowledge.

One important area of knowledge came from the Persians. They made discoveries about water beneath the ground. Such water is called **groundwater**. It is water that has seeped through the soil and flows slowly beneath the ground surface. Groundwater is different than a river. You cannot see groundwater until it comes to the surface as a spring.

One Persian scientist named Mohammed Karaji lived during this time. He wrote a book called *The Extraction of Hidden Waters*. Extraction means "pulling out." His book was especially important because of where he lived. The Middle East does not have a great deal of water that is easy to reach. Knowing how to find water in such areas is very valuable knowledge.

The Mustansiriyah School in Iraq is the first-known university in the world. It was established early in the thirteenth century.

Karaji studied at learning centers like the Library of Hulwan in Baghdad.

Baghdad is now in modern-day Iraq.

Europeans learned of Karaji's techniques through trading centers like Enisala in Romania.

Karaji spent most of his life working in Baghdad. He was mainly a mathematician. He wrote many books on math topics. When he was much older, he needed to make some extra money. He decided to write and publish a book about water. This book shows that Karaji had a deep understanding of groundwater. In the western world, this knowledge wasn't gained for nearly 700 more years!

Karaji's book shows that he was familiar with the main ideas of the water cycle. In his book, he tells about each part of the cycle. He also shows a strong understanding about soil and the best places to find **freshwater**. He knew how water moves underground. He also invented new and brilliant ways to dig underground and find water. The methods he figured out are still in use in many parts of the world.

Karaji's book is the oldest known book on groundwater. The information in the book is mostly the same as what scientists know today to be true.

Karaji's methods are still used today.

Areas with groundwater are really layers of soil and rock through which water flows.

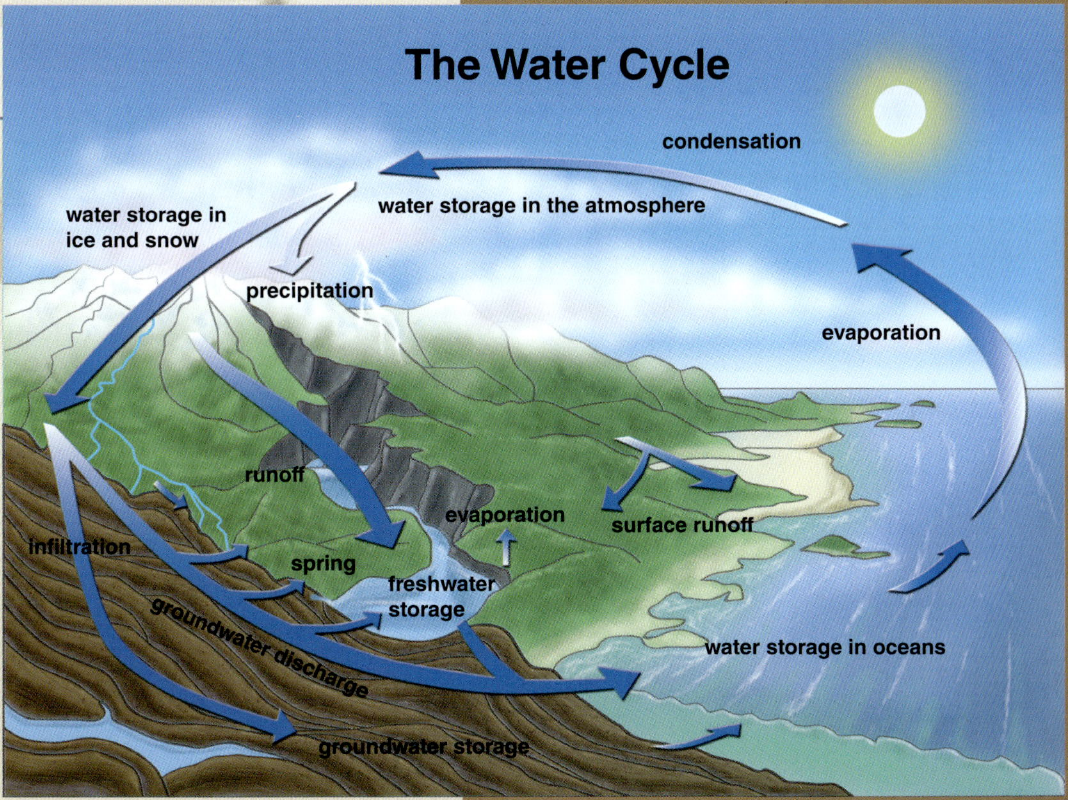

The Water Cycle

Water on Earth is always on the move. It's moving from one place to another and one form to another. Water is **evaporated** from the ocean and from the land. It forms clouds in the sky. It falls to Earth as rain, snow, sleet, and hail. It runs off into rivers, streams, and lakes. It seeps into the ground. Living things use it for many purposes. Eventually, the water flows back to the ocean and the process starts again. This is called the water cycle.

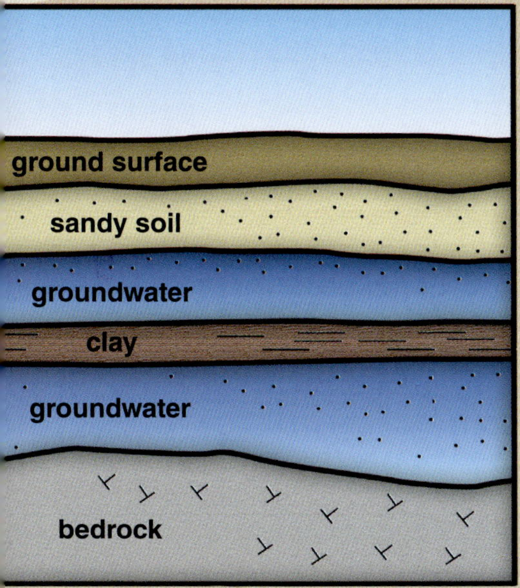

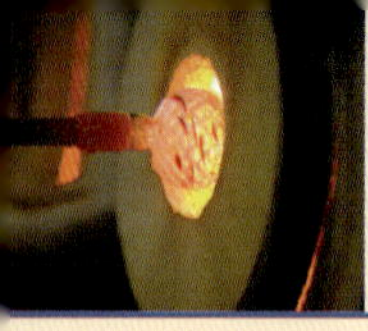

Bernard Palissy

Bernard Palissy was born in France around 1509. He worked mainly as an artist. During the time when he lived, it was common for people to learn more than one type of work. Palissy worked as a scientist, too.

When he was young, Palissy was trained in many things. He learned to blow glass. He painted portraits. He was trained as a land surveyor, which is someone who figures out the lay of the land both on and below the surface. He surveyed the salt marshes in a local town.

At one point in his life, Palissy was shown a piece of a cup. It was probably made of **porcelain** from China. Porcelain is a type of very fine pottery. Palissy was struck by its beauty. He wanted to learn how to make artwork just like it.

Pottery is placed in a red-hot kiln to turn the clay into useful items.

For many years, Palissy tried to discover the secrets of making such dishware. He never succeeded in making that particular style. However, he did learn a great deal about the earth. He couldn't help it. He spent so much time looking for materials to make pottery that he got to know the earth very well.

Even though he didn't create porcelain, his work did capture the attention of the royalty in Paris. They gave him money and a place to work in the special area of Paris where tiles were made.

Land surveying is a profession that goes back through all recorded history. The same rules of math and science that were used long ago are still used today.

Palissy put animals and plants on his works.

Because Palissy had a lot of knowledge about the earth, he began to teach others about it. He gave many talks about natural history. They included his ideas about water springs, groundwater, and fossils. His ideas were far in advance of what was known at the time.

During his life, Palissy wrote at least two major works. They tell about many science topics. The second book is the most detailed. It connects the flow of rivers and streams to rainfall. It also talks about how water flows. In his books, Palissy described a theory that was new in the western world. We know it today as the water cycle.

Sadly, Palissy was a victim of his time. In those days where Palissy lived, when people had ideas about religion that were different from most other people, sometimes they went to prison for it. That is what happened to Palissy. He didn't follow the Roman Catholic Church, the main church of France. He was sent to the Bastille, a French prison. Sadly, he died there of poor food and bad treatment.

dinosaur bones fossilized

Scientist Ahead of His Time

Long before most people knew it, Palissy figured out how fossils are made.

Bastille prison in France

This chemist is testing groundwater for contamination by chemicals used in farming, which make it unsafe to drink.

Cinda Crabbe MacKinnon

Today, water scientists are needed to be sure our water is clean and safe for living things to use. Cinda Crabbe MacKinnon started a company that tests the soil and groundwater for harmful chemicals. If there is a problem, her company tells what to do about it. Her company is just one of many that do this kind of work. Without them, our life-giving water might be deadly instead. MacKinnon says, "Geology is a challenge. It's like finding clues to a mystery and working out the answer." McKinnon keeps the mystery out of our water!

Edmond Halley

Edmond Halley was born in 1656 in England. His father was a wealthy businessman. Halley did not have to go to work right away. He was able to study and get a good education.

Halley had a great deal of interest in many areas of science. He was most interested in **astronomy**. That is the study of space and the objects in it. Today, Halley is best known for the comet that is named after him. It is Halley's Comet. He studied comets in school and figured out that a few comets people wrote about were really just one comet. He used math to show the orbit of the comet. The comet was named after him. You can see it in the sky from Earth every 76 years.

Comets appear as long, bright streamers in the sky.

Earth to Sun

Halley figured out a way to measure the distance from Earth to the sun. But it could only be done two times every 100 years. Those times did not happen during Halley's lifetime. He left the information to other scientists. At the right time, scientists around the world used his plan to figure it out. It worked!

For a period of a year and a half, Halley lived on the island of St. Helena in the Atlantic Ocean. He spent the time studying the position of stars. He charted the position of 341 stars during this time. This was especially amazing because of the poor weather and conditions on the island. This accomplishment made Halley famous among scientists. It is also very possible that the time spent on the island taught Halley a great deal about **meteorology**. That is the study of weather. And understanding weather is an important part of understanding water.

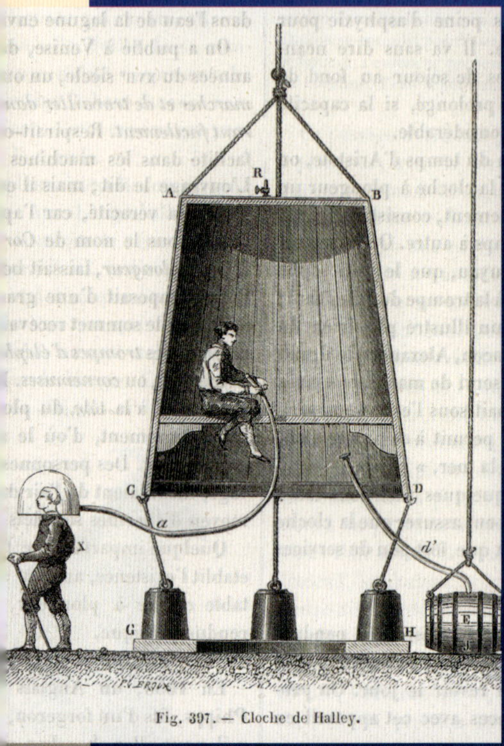

Diving Bell

Halley made plans for a diving bell. A diving bell is something like a submarine. A scientist can go inside the bell for a long period of time and sink down below the surface of the water to study water and life in the sea. This idea was far ahead of anything being planned or used at the time.

Halley's work in water science relates mainly to weather. In 1686, Halley published an important paper and a chart on the **trade winds** of the earth and **monsoons**. Trade winds are winds that blow mainly east to west in regular patterns. Monsoons are big, violent rainstorms. In this same paper, Halley wrote that the sun is the driving force behind most of the weather on Earth. He was right. He also showed the relationship between air pressure, **altitude** (height above sea level), and weather. Air pressure and altitude affect the weather.

Halley also thought up important ideas about where the **minerals** in the oceans come from. Minerals are basic chemicals that are found in rocks and soil. Halley realized that when rain flows over land, it dissolves minerals from the land. As the water flows, it carries the minerals to the oceans and seas.

Halley lived a long life and accomplished a great deal with his work. He died in 1742.

George Hadley

Halley's theory on Earth's trade winds couldn't explain why they always blew from east to west. George Hadley lived about the same time as Halley. He was a lawyer and amateur scientist. He came up with a more complete explanation of why trade winds happen. He explained that the sun would evaporate a great deal of water near the **equator**. The water vapor in the warmer air would rise up into the atmosphere and flow north and south. This water travels a long distance in the atmosphere and cools. Because it cools, it falls back to Earth. The falling rain pushes air down. Wind is caused by this motion of the air falling. The way Earth rotates on its axis causes the wind to blow east and west. These east-to-west winds are called trade winds.

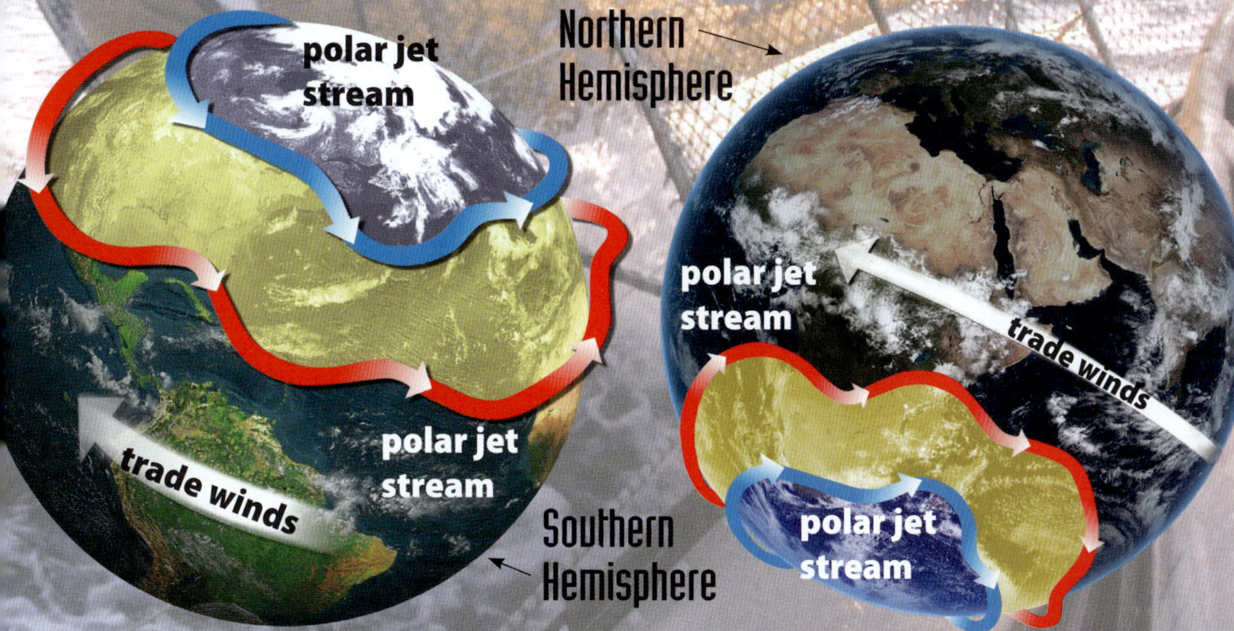

Henry Darcy

Henry Darcy was born in Dijon, France, in 1803. His father died when he was just 14 years old. His mother borrowed money to keep him in school. He studied at science and **engineering** schools. He was an honor student. He graduated high in his class.

After graduation, he joined the French Corps of Bridges and Roads. That was the government group in charge of building such structures.

cathedral in Dijon

Because the reservoir was at a higher altitude than the city, Darcy could build fountains that used gravity to shoot water into the air. The people of Dijon drew water from fountains like this.

Darcy devoted his work to the corps. He spent most of his life working in his home city. He quickly became the corps' lead engineer for that whole area. As the lead engineer, he was able to design a water system that gave water to the people in his town.

Before that time, people hauled their water from rivers and streams to their homes. Darcy figured out a way to bring water right into the city for everyone's use. The people could go to one of many water posts around town to get their water. They were very grateful to Darcy. He became a hero to them. Dijon was one of the first cities in Europe to have such a system. The country of France gave him an important medal because of his work.

Darcy's system included pipes and the use of gravity to make water flow into the city from a **reservoir** about 12 kilometers (7.5 miles) away. The system included about 28,000 meters (92,000 feet) of pipes. It was especially amazing because it needed no pumps or filters.

As a water engineer, Darcy studied many other important things about the way water behaves in water systems. These include such things as pipes and **channels**. One of the most important things that Darcy studied was the way that water flows through sand. He led many experiments with his coworkers. Based on these experiments, Darcy came up with an important equation. It is called Darcy's Law. Scientists still use it to help get water from underground.

These experiments and his report about them were the last work Darcy did. He died when he was only 55 years old. Soon after, the people of Dijon built a monument to him.

Darcy was a hero to the people of Dijon.

Carol Browner

Carol Browner grew up near the Florida Everglades. Her parents taught her the importance of caring for the earth. She grew up to become a lawyer. She wanted to work to protect our land, water, and air. When she was just 38, the U.S. president made her the head of the Environmental Protection Agency (EPA). She held the job for eight years, the longest anyone ever has. A big part of her work was to protect and clean our water. During her time in office, she made sure that millions more people than ever before had clean water to drink and use in their houses.

Pelageia Polubarinova-Kochina

Pelageia Polubarinova was born in Russia in 1899. Her parents both valued education. The family moved from her hometown to St. Petersburg, the capital city of Russia, when she was young. They did this to get a better education for their children.

Polubarinova-Kochina graduated from high school in 1916 and entered the university. Sadly, her father died in 1918. She had to get a job to continue her education. She was very smart. She was able to get a job at an important science lab.

While at the university, things became more difficult for her and her sister. They both got a terrible disease called tuberculosis. It is a disease of the lungs. Her sister died of the disease. Polubarinova-Kochina recovered and was able to finish school. She earned a degree in math.

In 1925, she married Nikolai Kochin. She had met him at the university. They had two daughters. Polubarinova-Kochina left work at the lab to raise her daughters. She still continued her research. She also taught at several schools.

These times in Russia were very troubling. There was revolution in her country. There were two world wars. Her husband became ill and died during the second war. He was in the middle of teaching classes at the university when he died.

a street, canal, and cathedral in St. Petersburg

Polubarinova-Kochina took over her husband's work. She also began to hold important jobs in her country. She led a great deal of major research. A big part of her study had to do with using math to understand the flow of liquids, especially water. She developed important ideas about how water flows underground. She also developed important ideas about how water flows through filters and is cleaned.

Understanding the flow of water is a big part of understanding water. It also makes it easier to get water from one place and move it to another. Polubarinova-Kochina's research and writing helped to explain a lot about the way water flows.

She wrote many books about both the things she studied and the important scientists she studied and worked with. One of these scientists was her husband. These books are used by scientists all over the world.

Polubarinova-Kochina received many awards during her life. When she was 95, a special conference among countries around the world was held. It honored her birthday, and she gave the opening talk. She worked all the way until her death at the age of 100.

In fact, she was 100 when she published her last paper. She wrote it with her daughter, who was then in her seventies.

Pelageia Polubarinova-Kochina had a long, interesting, and important life. She lived through many tough times during her life. She made important contributions to science and our understanding of water. Her work will live on.

cross-section of a water filter when first purchased (left), and after six weeks of use (right)

waste water treatment facility

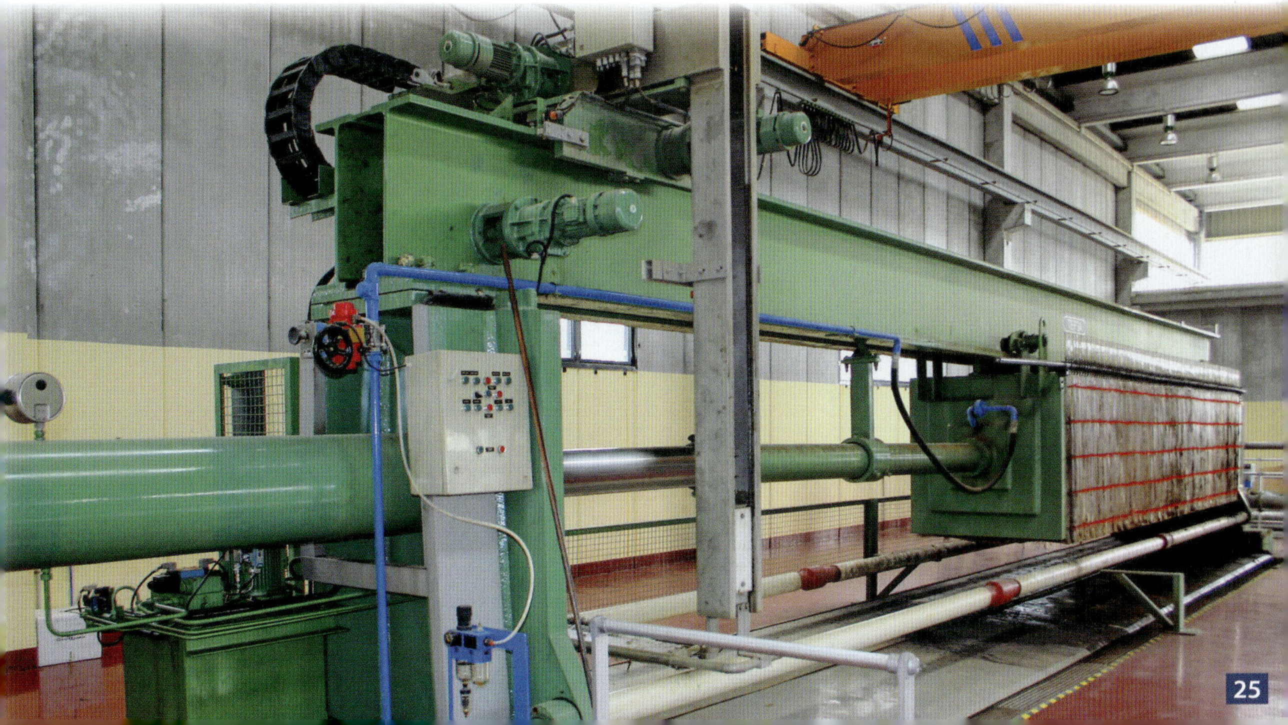

Environmental Engineer: *Alexandria Boehm*

Stanford University

Surfing Scientist

Ali Boehm is never far from water. She plays in the water *and* she works in the water. She grew up surrounded by it—in Hawaii. She loved to snorkel, scuba dive, and surf.

But the places she swam started filling with pollution. Boehm wanted to do something about it. "I've been an environmentalist since I was really young," she tells Sally Ride Science.

In college she wanted to study the environment. And she also wanted to solve problems. She decided environmental engineering was a good way to do both.

"Nowadays, I take a lot of water samples with my students," she says. "Being outdoors and doing science is what I enjoy the most." But she also takes time to ride the waves. Hang ten!

Being There

"I miss Hawaii a lot, especially because my family lives there. I also miss the weather, and the beach, and the ocean."

Boehm gets samples of ocean water. She takes them back to test them in her lab.

Is It 4 U?

If you were an environmental engineer, you might . . .

- find out how to clean up oil spills.
- prevent soil erosion.
- design buildings that don't waste energy.

Boehm tests the ocean water for bacteria that can make swimmers sick.

Lab: *How Water Soaks into the Earth*

When it rains, water soaks into the earth. The size of the particles in the soil affects how fast the water soaks in. Why does this matter? More water can soak in faster with certain kinds of soil. This helps the water supply for the people in the area. It can also make certain areas less likely to flood.

Materials

- water
- measuring cup
- coffee can or similar container
- sand
- unpopped popcorn
- pebbles or gravel (about the size of marbles)
- thin tool to make a hole in the can
- stopwatch or a watch with a second hand
- paper and pencil

Procedure

1 Make a hole in the side of the can near the bottom. The hole should be about 2 to 3 millimeters ($\frac{1}{8}$ to $\frac{1}{4}$ inch) big. Be careful! Have an adult help with this.

2 Cover the hole with your finger. Fill the can with sand up to 2.5 cm (1 inch) from the top.

3 Fill the measuring cup with water. Slowly pour it over the sand so that it soaks in. Keep adding water until it doesn't soak into the sand anymore. Important: Keep track of the amount of water you pour into the sand.

4 Get your watch ready. Remove your finger from the hole and start timing. (You may want to place the can in the sink or on the sink's edge so the water flows into it.) Stop timing when there is about one drip of water per second coming out of the can. Record the result in a chart.

5 Empty the contents of the can. Wash the can well with water so there is no sand left.

6 Repeat steps 2–5 with the popcorn and then with the pebbles or gravel. For each, use the same amount of water as with the sand. Don't worry about the water soaking in.

7 Compare the results. Through which substance does the water flow fastest? Slowest?

Extension

To extend this activity, measure water flow through each substance three times in a row. Compare those results. Is there a big difference between the water's flow the first time through the substance and the second and third times? Also, you can catch the water that runs out and measure it to compare its volume. Why doesn't all of the water flow out from each substance?

Glossary

altitude—the height of an object above a reference level, especially above sea level or the earth's surface

astronomy—the scientific study of the universe and of objects that exist naturally in space, such as the moon, sun, planets, and stars

atom—the smallest unit of an element that can exist alone or in combination with other elements

channels—a passage for water or other liquids to flow along, or a part of a river or other area of water that is deep and wide enough to provide a route for ships to travel along

element—a pure chemical substance that cannot be broken down into anything simpler by chemical means; the fundamental materials of which all matter is made

engineering—the work of an engineer (someone who designs and builds something using scientific principles)

equator—an imaginary line drawn around the middle of the earth an equal distance from the North Pole and the South Pole

evaporate—to cause a liquid to change to a gas, especially by heating

freshwater—of, relating to, living in, or consisting of water that is not salty

groundwater—water beneath the earth's surface that supplies wells and springs and is often found between saturated soil and rock

meteorology—the study of processes in Earth's atmosphere that cause weather conditions

mineral—a chemical substance that is formed naturally in the ground

molecule—the simplest unit of a chemical substance, usually a group of two or more atoms

monsoon—the season of heavy wind and rain during the summer in hot Asian countries

porcelain—a hard but delicate shiny white substance made by heating a special type of clay to a high temperature; fine china

reservoir—a natural or artificial pond or lake used to store water, control floods, or generate electricity; a body of water stored for public use

trade wind—a wind that blows mainly east to west in regular patterns

water cycle—the cycle of evaporation and condensation that controls the distribution of the earth's water as it evaporates from bodies of water, condenses, precipitates, and returns to those bodies of water

Index

Sally Ride Science

Sally Ride Science™ is an innovative content company dedicated to fueling young people's interests in science. Our publications and programs provide opportunities for students and teachers to explore the captivating world of science—from astrobiology to zoology. We bring science to life and show young people that science is creative, collaborative, fascinating, and fun.

Image Credits

Cover: Nik Niklz/Shutterstock; p.3 Tomasz Szymanski/Shutterstock; p.4 (top) Melanie Jensen/Shutterstock; p.4 (bottom) Photos.com; p.5 Tim Bradley; p.5 (right) VIKTOR DRACHEV/AFP/Getty Images; p.6 (top) Ian Wilson/Dreamstime.com; p.6 (bottom) AP; p.7 (top) The Granger Collection, New York; p.7 The University of Texas at Austin; p.7 (bottom) Julius Costache/Dreamstime.com; p.8 Jesper Jensen/Alamy; p.8–9 Tim Bradley; p.8–9 (background) Joaquin Ayllon Perez/Shutterstock; p.9 Tim Bradley; p.10 (top) Shutterstock; p.10 (left) Mary Evans Picture Library/Alamy; p.10 (bottom) iStockphoto; p.10–11 CORBIS; p.11 Brian McEntire/Shutterstock; p.12 (left) Ng Yin Chern/Shutterstock; p.12 (right) The Granger Collection, New York; p.13 SPL/Photo Researchers; p.14 (top) Varina and Jay Patel/Shutterstock; p.14 (left) North Wind Picture Archives/Alamy; p.14–15 Andrew F. Kazmierski/Shutterstock; p.15 NASA; p.16 Mary Evans Picture Library/Alamy; p.16–17 Jefras/Shutterstock; p.17 (bottom) Tim Bradley; p.18 (top) Yan Simkin/Shutterstock; p.18 (left) Tibor Bognar/Alamy; p.18 (right) Courtesy of Henry Darcy Family; p.18–19 Tim Bradley; p.20 (left) Courtesy of Glenn Brown; p.20 (right) Courtesy of Glenn Brown; p.21 (top) Terry Ashe/TIME Life Pictures/Getty Images; p.21 (background) Tomasz Szymanski/Shutterstock; p.22 (top) Brian Tan/Shutterstock; p.22 (left) Rick Reason; p.22–23 Zastavkin/Shutterstock; p.24 Brian Tan/Shutterstock; p.25 (top) Dr. Jeremy Burgess/Photo Researchers, Inc.; p.25 (bottom) Joan Ramon Mendo Escoda/Shutterstock; p.26 Stanford University; p.27 US Dept. Agric. Susan Boyer; p.27 (bottom) Phil Degginger/Getty Images; p.28 (top) Tony Wear/Shutterstock; p.28–29 Nicoll Rager Fuller